At the Construction Site

Published by The Child's World®, Inc.

Design and Production:
The Creative Spark, San Juan Capistrano, CA

Photos: © 1998 David M. Budd Photography

Library of Congress Cataloging-in-Publication Data
Sirimarco, Elizabeth 1966-
    At the construction site / by Elizabeth Sirimarco.
        p.   cm.
    Includes index.
    Summary: Briefly describes what goes on at a construction site and
some of the equipment used there.
    ISBN 1-56766-573-X (lib. bdg. : alk. paper)
    1. Building sites Juvenile literature. [1. Building. 2. Construction equipment.] I. Title.
TH375.S57   1999
690—dc21   99-10829
                                                                                              CIP

At the Construction Site

Written by Elizabeth Sirimarco
Photos by David M. Budd

F I E L D   T R I P S

The Child's World®, Inc.

Wow! What are they building over there?

There is a lot of noise at this **construction site.**

Clang! Crash! Bang!

A construction site can be dangerous.

Sometimes heavy objects fall from above!

Workers wear hard hats to protect their heads.

Hard hats are made from strong, heavy plastic.

A worker looks at a big sheet of paper.

He calls it a **blueprint.**

It shows the workers how to **construct**
the building.

What are they building here?

Look at that giant **drill!**

It spins around and around. Whirrrrrrrrr!

It makes a big hole in the ground.

The workers will pour **concrete** into the hole.

The concrete will dry to form a post.

The post will help make the building strong.

9

A **crane** is the tallest machine at the site.

It is shaped like the letter "T."

On a construction site, a crane lifts many things.

It can move heavy objects.

The crane's **tower** holds a long arm.

The arm is called a **jib.**

The tower stands still, but the jib moves.

The crane has a cab.

There are **controls** in the cab.

A worker uses the controls to move the jib.

The cab is so high! To reach the cab, the worker climbs a long, long ladder.

The ladder is in the middle of the tower.

Can I drive?

A crane has long metal **cables.**

The cables can move a heavy object.

A worker attaches the cables to the object.

Sometimes a crane picks up
something small.

Usually a crane picks up heavy objects.

This is a steel **form.**

The workers pour concrete into the form
to make a **girder.**

Careful!
Don't drop it!

19

Here comes the concrete mixer!

The barrel on the back of the truck spins around and around.

It mixes sand, water, and a special gray powder to make concrete.

Cement is the powder used to make concrete. It helps the sand and water stick together.

The concrete mixer cannot drive on the construction site.

The workers must move the concrete to where they need it.

They pour the concrete into a **pump.**

Slop! Slop! The concrete is wet and cold.

The pump has a very long hose.

Concrete travels through the hose.

The hose is attached to a big, orange arm.

The arm moves the concrete to where the workers need it.

One worker holds the hose steady.

The others will spread the concrete.

They make sure it is smooth.

Workers wear tall rubber
boots to walk in wet concrete.

Another machine can dig and move dirt.

A claw at one end digs holes.

There is a scoop at the other end.

It picks up a big load of dirt.

Scoop! Scoop!

It dumps the load into a narrow space.

Look at this drawing.

It is a picture of what the building will

look like someday.

What a pretty place!

Let's come back next week
to see what happens next!

31

Glossary

blueprint (BLEW-print) — A blueprint is a drawing that shows construction workers how to make a building. It is called a blueprint because the drawing is bright blue.

cables (KAY-bulls) — Cables are strong ropes, often made from metal. Workers attach a crane's cables to an object that they want to lift.

concrete (KON-creet) — Concrete is a wet material that gets hard and strong when it dries. Construction workers use concrete to make many things, such as buildings and roads.

construct (kun-STRUKT) — When you construct something, you build it. Workers construct a building.

construction site (kun-STRUCK-shun SITE) — A construction site is a place where workers build something. The workers make a building at the construction site.

controls (kun-TROLZ) — Controls are used to guide a machine. A crane's controls help the workers move the jib or lift an object.

crane (KRAYN) — A crane is a large machine. Construction workers use a crane to lift and move heavy objects.

drill (DRIL) — A drill is a tool used to make a hole in something hard. Construction workers may use a drill to make a hole in the ground.

form (FORM) — A form is something used to shape an object. If a worker pours concrete into a form, it will dry in the same shape as the form.

girder (GIR-dur) — A girder is a long, sturdy pole that helps support a building. A girder is usually made of metal, concrete, or wood.

jib (JIB) — The jib is part of a crane. It is a long arm that lifts and moves an object.

pump (PUMP) — A pump is a device that lifts and moves something wet. A construction worker may use a pump to move wet concrete.

tower (TAU-er) — A tower is the tall part of a crane that sits on the ground and rises high into the air. The tower supports a long arm that lifts heavy objects.

Index

Elizabeth Sirimarco has written more than 20 books for young readers. She and her husband, photographer David Budd, live in Colorado.